T5-BBY-702

IMPLICATIONS OF THE 1976 ARAB-ISRAELI MILITARY STATUS

Robert J. Pranger
Dale R. Tahtinen

American Enterprise Institute for Public Policy Research
Washington, D. C.

Robert J. Pranger and Dale R. Tahtinen are, respectively, director and assistant director of foreign and defense policy studies at the American Enterprise Institute for Public Policy Research.

ISBN 0-8447-3209-5

Foreign Affairs Studies No. 34, April 1976

Library of Congress Catalog Card No. 76-8381

Printed in the United States of America

CONTENTS

IMPLICATIONS OF THE 1976 ARAB-ISRAELI MILITARY STATUS

1. TARGETS AND WEAPONS IN THE NEXT ROUND OF MIDDLE EAST WAR

Robert J. Pranger

Four zones of possible warfare exist in the Middle East. While they can be distinguished according to the targets involved, there is considerable overlap. In zone 1 are frontline battlefield targets, including surface-to-air missile (SAM) umbrellas in their immediate vicinity. During the Egyptian move across the Suez Canal in October 1973 a screen of Russian-supplied SAM-6s, SAM-7s and ZSU-23s was established on the east side of the canal, to provide a forward wall of anti-aircraft defense. Such mobile missiles and anti-aircraft batteries would obviously be in zone 1. At the same time, however, Egypt had its well-developed SAM systems on the west bank of the Suez Canal to protect ground operations in the Sinai as well as to bar entry of Israeli military planes into Egyptian airspace. Before October 1973 the SAM belt west of the Canal would also have been in zone 1, but as a result of the disengagements attendant on the interim peace accords these sites are far enough to the rear to qualify as zone 2 targets in any hostilities (see below).

Zone 2 targets embrace all military rear-echelon strong points such as airfields, supply stations, troop staging areas, command centers, the SAM systems discussed above and those guarding important installations, and so on. Targets in this zone could be found all over a given country.

Economic, communications, and transportation infrastructure would comprise the zone 3 targets. Included would be large targets such as dams, and small installations such as microwave relay stations.

Such facilities would extend throughout a nation, and even into its civilian population centers. Rail yards and ports, for example, are usually adjacent to neighborhoods where people carry on their daily life—neighborhoods that include schools, hospitals, factories, and recreation areas.

Zone 4 targets are the purely civilian populations. Attacks aimed at zone 4 targets would be designed to create widespread terror, much in the way strategic bombing in World War II often operated, and in the manner in which so-called "countervalue" strategic nuclear war might be waged between the United States and the Soviet Union.

Military Planning in the Middle East for a Future War. Looking at the kinds of new military equipment increasingly being favored by Israel, and presumably to be favored by the Arabs, one may conclude that the principal emphasis in Middle East war planning is to be on surface warfare by forces not directly engaged in front-line combat. Such warfare would be directed at targets in zones 2 and 3, and possibly zone 4, and would be waged by air, land, and sea forces firing on surface (land) targets. In a sense, this kind of planning for surface warfare has been favored by Israel ever since the preparations began for the June 1967 conflict that opened with thorough Israeli demolition of Egypt's air force as it stood exposed on the ground. At the time, this approach was commonly called "preemptive." During the 1969–1970 war of attrition Israel's favorite response to Egyptian artillery was "deep penetration" into Egypt proper. In the past, Israeli reprisals for guerrilla attacks have leaned toward retaliatory air strikes inside Jordan and Lebanon. While both Egypt and Israel now favor use of the word "preventive" to describe this military leaning, the basic idea is the same as the basic idea behind preemption: strike first at the enemy behind the battlefield, thus cutting off his air cover, supply lines, and home front (the last by psychological impact as much as by real destruction). During the October 1973 war Israel was inhibited in this strategy by effective and diverse Egyptian and Syrian air defense systems over the battlefield area, and in zone 2 by dispersal of airfields and better protection of aircraft. Israeli air power did escalate the Syrian fighting by moving against zone 3 infrastructure targets, however, with the express purpose of demoralizing Syria on the battlefield. And air attacks were staged against Egyptian airfields, according to Israeli sources (denied by Egypt).[1]

---

[1] "Israel Seeks Preemptive Strike Capability," *Aviation Week & Space Technology*, 27 October 1975, p. 19. A report based on remarks by senior Israeli military officials at a symposium on the October 1973 war in Jerusalem during October 1975.

Table 1

WEAPONS AND TARGETS IN HYPOTHETICAL BATTLE ZONES

Weapons	Zone 1 Military/ Battlefield	Zone 2 Military/ Rear	Zone 3 Infrastructure	Zone 4 Civilian
Aircraft	X	X	X	X
Air-to-air missiles	X			
Surface-to-air missiles	X			
Air-to-surface missiles	X	X	X	X
Surface-to-surface missiles	X	X	X	X
Armor	X			
Anti-armor	X			
Helicopters	X	X	X	X
New weapons	X	X	X	X
Nuclear	X	X	X	X

Note: Xs represent weapons applicability.

One may presume that high on the list of Israeli weapons acquisitions and requests since October 1973 have been arms for effective preventive strikes against rear-echelon targets in Egypt and Syria should another round of fighting threaten. Israel's defense officials have intimated as much.[2] Table 1 shows the varieties and primary missions of the major weapons systems and the varieties of military targets in the Middle East today. The systems could, of course, be used in other zones as well as those for which they are primarily designed (for example, armor and anti-armor might break through front lines and operate in rear-echelon areas as in the October 1973 war). If one looks at this diagram and at recent Israeli and Arab acquisitions one can appreciate how heavily the scales are weighted toward warfare in zones 2 and 3 in any future fighting.

[2] Ibid.

Tables A-4 and A-6 in this study (major surface-to-surface and air-to-surface missiles) are especially relevant. To these lists one should add laser-guided bombs and drones: weapons such as these can be launched from ground, air, and naval equipment. In future fighting one should not discount the possibility of naval bombardments, using surface-to-surface missiles, against land-based targets.

One can, of course, argue that the simple presence of capabilities does not augur actual intentions. Moreover, the laments of military leaders about a past failure on Israel's part to use preemptive options should not be taken as proof of future intent. Yet it cannot be overemphasized that, in fact, the United States and (probably) the Soviet Union know very little about the full military intentions of the Arabs and Israelis. Some American ignorance is obviously the result of poor (or nonexistent) intelligence; it is probable that, before October 1973 and even to this day, the United States has had little or no independent surveillance of Israel's defense planning, and this despite the fact that the United States was caught wholly unprepared for Israel's surprise attack in June 1967. The Middle East wars of 1956, 1967 and 1973 were planned in utmost secrecy, so that even with superior intelligence it might have proved impossible to penetrate preparatory activities. During 1973 the Egyptians and Syrians resorted to excellent camouflage of their intentions.[3] Similarly, despite close U.S. intelligence collaboration with Israel in a number of areas, it seems unlikely that U.S. officials catch much glimpse of Israel's defense plans. This is not to prejudge the question whether the United States should have detailed knowledge of Israeli plans. A familiar argument against this foreknowledge is that it would establish prima facie evidence for American collusion with Israel in an Arab world already predisposed to believe that such collusion exists. This "what-you-don't-know-won't-hurt-you" attitude must be balanced by a contrary consideration that surprise may eventually trigger a U.S.-U.S.S.R

---

[3] The Insight Team of the London *Sunday Times, The Yom Kippur War* (Garden City, N.Y.: Doubleday, 1974), pp. 101-111. Soviet knowledge and possible complicity remains mysterious. Public literature, including the exhaustive account cited above, offers some hints, but no explicit well-documented record of Moscow's role in planning the October 1973 war exists. On one hand there is Israeli General Chaim Herzog's assertion that, "[t]here are many indications that in the third week of September 1973 the top echelon in the Soviet Union was fully aware of the Egyptian plan to go to war"; see Herzog, *The War of Atonement, October 1973* (Boston: Little, Brown, 1975), p. 287. On the other hand there is the statement of Soviet Ambassador to Egypt Vinogradov in April 1974 that President Sadat had given the Russians two days' notice to leave Egypt; see Vladimir Petrov, *U.S.-Soviet Détente: Past and Future* (Washington, D. C.: American Enterprise Institute, 1975), p. 32, n.3.

nuclear confrontation or another oil embargo (or both) with consequences far beyond exposure by Arabs who are already disposed to believe the worst in any case.

It is surprising how far in advance war plans in the Middle East can be fully developed and still remain undetected by outsiders. Israel and France worked closely for two years before the 1956 Middle East war, with their collaboration including extensive secret French arms shipments to the Israelis in clear violation of the 1950 tripartite arms control agreement among France, Great Britain, and the United States.[4] Apparently, detailed plans were completed by Israel nearly one year before the June 1967 war.[5] Egyptian-Syrian planning for the October 1973 conflict was concluded much closer to actual attack than was the planning in the two Israeli cases, but President Anwar Sadat gave the clear signal to move on such plans as early as 14 November 1972 in a speech to a closed meeting of Egypt's Socialist Union central committee.[6] In all three wars the actual date of attack came as a surprise to the United States—Israeli paratroops near the east end of the Mitla Pass on 29 October 1956, Israel's Mystère IVs and supersonic Mirages against Egyptian airfields on 6 June 1967, and Egyptian infantry and Syrian armor against Israeli complacency on 6 October 1973. The next round of fighting in the Middle East might likewise begin as a surprise, not only because American intelligence has surveillance weaknesses, but because defense planning in the region is highly secretive in the first place. Of course, such secretiveness is obviously typical of military plans in all nations, including the United States, but it seems to have been particularly strong in the Middle East.

The point of this digression into secret war planning in the Middle East is not to cast blame, but to underscore a point: for the most sensitive of military matters, among which preemptive or preventive war plans would be foremost, it is well-nigh impossible to divine the intentions of the combatants either overtly or covertly. As former Secretary of Defense James Schlesinger noted in his fiscal year 1975 defense report (referring to the Russians), where knowledge of

---

[4] Kennett Love, *Suez: The Twice-Fought War* (New York: McGraw-Hill, 1969), pp. 433-442; Moshe Dayan, *Diary of the Sinai Campaign* (New York: Harper & Row, 1966), pp. 29, 31-32.

[5] Love, *Suez: The Twice-Fought War*, p. 677; Peter Young, *The Israeli Campaign 1967* (London: William Kimber, 1967), p. 87 (with reference to Brigadier General Ezer Weizman); Randolph S. and Winston S. Churchill, *The Six Day War* (Boston: Houghton Mifflin, 1967), p. 91 (with reference to Brigadier General Mordecai Hod).

[6] *The Yom Kippur War*, p. 59.

intentions is unclear, prudent military planning must judge from an opponent's capabilities.[7] While it is too easy and indeed somewhat misleading to infer intentions only from capabilities, it is probably safer to do this than to do nothing at all. Capability does say something about capacity, and capacity does determine the actual range of realistic military options. The main task of good intelligence analysis is to pinpoint the most likely choice of action in the given range. One of the chief failures of U.S. intelligence before the October 1973 war was to underrate Arab capabilities (and hence capacities), and thus to misunderstand Arab intentions. Having been warned by the Arabs on a number of occasions during 1973 that they intended to go to war against Israel, U.S. analysts did not pay attention because they underestimated Egyptian and Syrian capabilities, despite knowledge that new Russian equipment had arrived in 1973 and the possibility that Egyptian and Syrian forces had been training for the very kind of offense they staged on October 6.[8] Israeli assessments were at the base of U.S. inattention, since Israel's intelligence could not believe that there could be a successful Arab military effort, and U.S. assessments were heavily dependent on Israel's intelligence.

In 1976 the military capabilities of both the Arabs and the Israelis vastly outstrip the capabilities of each before October 1973. In particular Israel has increased its capacity to wage an effective preventive war against Egypt and Syria in zones 2 and 3, having a wide variety of missiles (including stand-off weapons and precision-guided munitions, called PGMs), guided bombs, and drones. Israel has advanced its capability despite improvements in Egyptian and Syrian air defense systems and anti-tank forces. Israel's defense forces might well become the first to exploit the full potential of new forms of ordnance against which defense is much harder than defense against any forms of ordnance in the past. The Soviet Union is now develop-

[7] *Annual Defense Department Report FY 1975* (Washington, D. C.: U.S. Government Printing Office, 1974), pp. 29-31.

[8] On U.S. knowledge of Soviet arms shipments to Egypt and Syria in 1973 see *The Yom Kippur War*, pp. 71-72; on American evidence of Sadat's military planning, ibid., p. 71. United States knowledge on training is less clear from the public record, but Egyptian forces were training on new Soviet equipment well behind Suez (ibid., pp. 59-60); Syrians were assisted in mid-1973 by the Soviet Air Force Commander and the Chief of the Czechoslovak General Staff (ibid., p. 72); noticeable preparations were under way by September 24 and apparently well-analyzed by the CIA (ibid., pp. 92-93); Egyptians were supplied with a new Soviet bridging device, the PMP bridge (ibid., p. 144); and Suez Canal engineers had used pressure jets for years to remove sand (ibid.), while Egypt's military had placed a big order for powerful, portable pumps in West Germany and received shipment in mid-1973 (ibid., p. 62).

ing a technological capability in PGMs similar to that of the United States, and the Arabs may be expected to press for a supply of these items unless suitable arms control arrangements can be negotiated among the superpowers (and possibly including France as well). Meanwhile, the Arabs may have to content themselves with such weapons as the French-made Martel, a tactical stand-off missile with a range of over thirty-five miles and capable of being launched from Mirage aircraft (see Table A-6).[9]

Where capability exists, there is the possibility for military action that would otherwise be sheer absurdity to contemplate. Israel's defense minister, Moshe Dayan, argued against an Israeli preemptive strike on Egypt just before 6 October 1973, even though signs of Egyptian military build-up were mounting. His grounds for rejecting this option were that Israel was technically unable to strike effectively at widely dispersed airfields defended by an ingenious anti-aircraft defense (though there may have been political reasons as well connected with possible U.S. displeasure over such an attack).[10] In other words, Israel did not have a preemptive capability sufficient for the realities of late 1973; even if it wanted to wage preventive war, its capacity to do so was severely limited. Since the preventive approach to its own defense was much favored by Israel's defense forces before the October war and found wanting on the eve of that war, one might plausibly assume that after the conflict Israel would seek to rebuild its preemptive capacities in light of the contemporary military realities in Egypt and Syria. In other words, the "lessons" of the October war may not necessarily have been in peaceful directions. This Israeli rebuilding would require forms of weaponry (most of them capable of conventional and nuclear possibilities) which could reach zone 2 and 3 targets with pinpoint accuracy and arrive largely unscathed by present air defense systems. Missiles fired from ground and sea platforms, stand-off weapons launched from aircraft outside the range of an air defense, guided bombs, and drones directed from air or

---

[9] Soviet PGM capabilities appear to be improving greatly, with the likelihood that such weaponry will eventually find its way into Arab arsenals. On PGMs in general and Soviet capabilities in particular, see: James Digby, *Precision-Guided Weapons*, Adelphi Paper No. 118 (London: The International Institute of Strategic Studies, 1975); Secretary of Defense Donald H. Rumsfeld, *Annual Defense Department Report FY 1977* (Washington, D. C.: U.S. Government Printing Office, 1976), p. 127; Drew Middleton, "Mass-Produced Precision Guided Weapons Are Said To Be Revolutionizing Military Doctrine and Tactics," *New York Times*, 23 February 1976.

[10] *The Yom Kippur War*, pp. 122-123. Some of Israel's military leaders were apparently more sanguine; see "Israel Seeks Preemptive Strike Capability."

ground launchers, all might accomplish hitherto impossible preemptive missions for Israel and possibly for the Arabs.[11]

From the Arab standpoint, the problem with Israeli weapons acquisition of this sort is as much that it improves Israel's capacity to wage preemptive war as it is that it signals actual intentions of doing so. In any event, it should be evident that the Arabs and Israelis are by no means expert in reading each other's intent. Prime Minister Yitzhak Rabin, among others, has said that Israel would not wage preventive war except under the most dire Arab provocation. This statement must give cold comfort in Cairo and Damascus.[12] For the purposes of establishing one's own national defense one must, if one errs, err on the side of overestimating one's opponents, especially in wartime, as Israel so painfully learned in 1973. In the midst of hostilities—and "no war, no peace" is a form of hostilities—prudence means leaning toward the worst case. No Arab leader could afford a repeat of Egypt's humiliation in 1967, as no Israeli leader could survive a repeat of 6 October 1973. In these circumstances, capacity must be matched by capacity, regardless of intentions—or so the logic of the Arab-Israeli struggle has dictated the situation in recent years. Double capacity to wage preemptive war in zones 2 and 3 with new and sophisticated military technologies is likely to lead to a highly unstable Middle East, because the chief defense issue will be how to anticipate a preemptive strike from forces that cannot be stopped or recalled once launched, and from missiles some of which, when launched, could be virtually on their targets as they were detected. To add to this instability, one cannot be sure whether such weapons would carry conventional or nuclear (or even chemical) ordnance, since most of them are capable of carrying a variety of explosive devices.

---

[11] A quantitative measure of amounts of new sophisticated weapons in the Middle East, such as PGMs, is impossible to compile from public sources. The dollar size of U.S. arms aid packages to Israel in the wake of the Sinai accords gives some idea of the overall magnitude—estimates have run as high as $15 billion in total commitments, by columnist Jack Anderson, while President Ford has referred to the total amount as "very substantial" without giving a dollar figure. See Bernard Gwertzman, "Ford Defends Israel Aid, but Denies Commitment," *New York Times*, 17 September 1975; Bernard Gwertzman, "U.S. Will Review Israeli Arms Need," *New York Times*, 16 September 1975; Jack Anderson, "The Peace Price: $15 Billion," *Washington Post*, 21 September 1975. The FY 1977 federal budget actually revealed a $500 million cut in arms aid for Israel. See *The Budget of the United States Government Fiscal Year 1977* (Washington, D. C.: U.S. Government Printing Office, 1976), Appendix, p. 79.

[12] On Sadat's own determination to have preventive war capability if it is vital for Egypt's defense, see Robert J. Pranger and Dale R. Tahtinen, *Nuclear Threat in the Middle East* (Washington, D. C.: American Enterprise Institute, 1975), p. 47, n.17.

Disengagement arrangements since October 1973 on both the Sinai and Golan fronts have probably been "purchased" by the United States with weapons that will have heavy impact on zones 2 and 3 should another round of fighting erupt between Israel and the Arab states.[13] Reports in the press have referred to the American early warning system as mainly a "symbolic" U.S. presence in the Sinai.[14] It is doubtful that many limitations have been placed by American policy makers on how such weapons might be used or what military strategies such items might encourage.

Six Hypotheses about the Next Round of Fighting in the Middle East. The targeting of zone 2 and 3 areas, with the possibility of a preemptive strike opening another round of warfare in the Middle East, can be called a "shift to the rear." From this shift one might draw certain hypotheses about the next Middle East war, should it take place, recognizing that hypotheses by definition are tentative assumptions open for further analysis and argument.

(1) *Given the large number of sophisticated arms poised for a strike against rear-echelon military and infrastructure targets in the Middle East, it is doubtful that air defense systems, no matter how good, can do more than wage limited war of attrition, with many enemy weapons piercing the defense.* The next war in the Middle East will not exact the heavy toll of Israeli aircraft exacted in the 1973 war by Arab surface-to-air missiles. Abundant missiles, guided

---

[13] Examples of the kind of sophisticated weaponry connected with the second Sinai disengagement agreement are found in "Israel to Receive Lance Missiles, F-15s," *Aviation Week & Space Technology*, 15 September 1975, p. 16. Also see the pledge in the Sinai accord, by the United States government, "to an early meeting to undertake a joint study of high technology, and sophisticated items, including the Pershing ground-to-ground missile with conventional warheads, with the view to giving a positive response." United States Senate, Committee on Foreign Relations, *Early Warning System in Sinai.* Hearings, 94th Congress, 1st session (Washington, D. C.: U.S. Government Printing Office, 1975), p. 253. Secretary Kissinger has denied that the Sinai disengagement agreement of 1 September 1975 was tied to an explicit list of weapons promised Israel, ibid., p. 242. The controversy which developed over the Pershing missile, sparked initially by columnist Jack Anderson's 16 September 1975 report of an agreement between the United States and Israel on this item, temporarily stalled Israeli acquisition of the weapon. Reports even began to circulate that the Pershing was of no special military significance to Israel, but only of psychological importance (a view that cannot be supported by the present analysis). However, recent reports indicate that Israel is pressing again for U.S. agreement to supply the Pershing. Drew Middleton, "Israel Again Asks for Pershing Missile," *New York Times*, 7 March 1976.

[14] Terence Smith, "In Sinai, Twang of Texas as Monitor Post Rises," *New York Times*, 14 February 1976.

bombs, and drones will in most cases obviate the need to have airplanes fly into SAM "envelopes" until such time as the SAMs themselves are destroyed by precision-guided stand-off weapons. This does not mean that mobile air defense in zone 1 (the battlefield) will not be important, as the Arab use of the SAM-6, SAM-7, and ZSU-23, as well as recent Israeli acquisition of the mobile U.S. Chaparral system, will attest. But SAM protection of zones 2 and 3 will be more problematical in the next round of fighting than it was in the last.

Under circumstances where strictly defensive measures fail, the only effective defense force would be one ready to stage its own preemptive action, should there be the remotest chance that the other side might attack first. He who "gets through first" could neutralize the other power, behind the battlefield and on the homefront (zones 2, 3, and 4).

(2) *The next round of fighting in the Middle East will involve efforts by one side or the other to deliver a preemptive "knock-out" blow to enemy forces in relatively quick order by focusing on rear-echelon military and infrastructure targets.* Some evidence exists that preemption is now considered an open possibility by Israeli defense planners.[15] Unfortunately, noncombatant advisers from foreign countries, especially the Soviet Union, are working in rear-echelon locations, and such persons will be involved, with little or no chance of advance warning or evacuation, if hostilities open with preemptive strikes against zones 2 and 3. Weapons promised to Israel by the United States under the same disengagement agreement whereby the American-manned Sinai early warning system was established include a number of items that will be targeted against rear echelons. Argument about the safety of Americans working at the Sinai station, a concern voiced during congressional debate on the warning system, misses the point that the chief danger to foreign noncombatants from the disengagement agreement will come from the arms supplied under its terms, and the foreign noncombatants affected will be personnel behind the lines, not personnel in the Sinai.

At the outbreak of new hostilities the United States and Soviet Union could quickly find that their citizens had become casualties, with responses by the superpowers unpredictable. Israel's raids into Egypt in 1969 and 1970 caused the Soviet Union to view with some alarm the threat to its advisers in Egypt, and Israeli attacks on a Russian transport ship in Syrian waters in 1973 brought ominous

---

[15] See "Israel Seeks Preemptive Strike Capability."

warnings from Moscow.[16] Foreign shipping may be more endangered in any new round of hostilities than it has been in the past, since it seems probable that naval warfare will increase in intensity.

(3) *Two preemptive capabilities poised for knock-out blows will undermine any concept of stable deterrence in the Middle East: the 1976 military status in the Arab-Israeli conflict is highly unstable and likely to become more so if current patterns of weapons acquisition continue.* The best defense against preemptive attack would be to anticipate the attack and stage one's own preemption. A second-best defensive strategy would be to let one's opponent know that one was poised for split-second reaction at the first sign of attack. Either way, such first-strike capabilities on both sides provide a hair-trigger for war in the Middle East: at the first indication of enemy attack, one must launch one's own attack or risk paralysis. That is to say, there will be launch-on-warning, even if the warning is a false alarm. The impact of preemptive weapons is to collapse distances in the Middle East, already short enough, to a size almost meaningless so far as advanced warning is concerned. (The U.S. Pershing missile has a speed of Mach-8, that is, approximately 100 miles per minute.) Not only are there no direct relations between the Arab states and Israel, thus ruling out "hotlines" between opposing military forces, but habits of mind in the Middle East tend toward surprise attacks, as the 1956, 1967, and 1973 wars demonstrated. Crisis management, without outside superpower intervention, seems to be hopelessly deficient. The combination of sophisticated weapons, preemptive strategies, short distances, exaggerated assessments, and minimal crisis-management machinery, suggests the possibility that unstable military situations may take on irrational dimensions: a first strike could be triggered out of fear that an enemy might attack in the context of some event not directly connected with the conflict itself (as, for example, a coup d'état).

(4) *The military situation in the Middle East, with deficient or nonexistent deterrence mechanisms, will be accident-prone—vulnerable to a wide range of mistaken judgments.* The hair trigger provided by a military situation where both sides possess first-strike

[16] A 1 June 1970 *Newsweek* story reported that Israeli attacks on Egyptian SAM-2 missile sites in 1969 resulted in killing 12 Russian advisers and wounding 29 others—see *Facts on File* (1970), 378 F3. During the October 1973 war Israeli missile boats sank three foreign freighters in the Syrian ports of Latakia and Tartus, including a Soviet ship. At an October 12 meeting of the U.N. Security Council the Soviet delegate called the sinking "barbarous"—see *Facts on File* (1973), 858 G1, A2.

capacities is made even more sensitive by knowledge that the enemy may attack mistakenly. Anticipation, and not the "early warning" systems now being installed in the Sinai, becomes the prime early warning technique. Not only is the American warning station in the Sinai largely symbolic, given its extremely limited functions, but it has been made irrelevant by the weapons provided under the same disengagement agreement that established the warning system in the first place. The United States seems simultaneously to have installed a warning system for the last war and provided weapons for the next. The only source of adequate warning in the highly unstable military situation in the Middle East would be an intelligence system that not only anticipated an enemy's planned surprise assault but even anticipated an opponent's mistaken attack in response to some innocuous signal from one's own military. There are prospects for an increasingly complicated instability under present arms acquisition policies in Israel and the Arab states. With the weapons now or soon to be available one can send false signals of intent as well as real ones, thus triggering a move by an opponent that would justify one's own preemption. It is fair to say that the evolving Middle East military situation has an unprecedented potential for destabilizing actions by Arabs and Israelis—actions that may be precipitated not only by cabinet decisions but by military groups acting (independent of formal policy-making processes) in order to force a preemptive attack. Indeed, the fear of a surprise move may become so acute that a change in an enemy's normal patterns of military activity, even if the change were innocent of any aggressive intent, could—if it were sufficiently sudden—precipitate a preventive thrust.

(5) *The best support for an unstable military position might appear to be a nuclear doomsday weapon in the Middle East.* That the system of anticipating mistaken as well as planned attacks can fail is self-evident. Its evident possibility for failure stems from the fact that both sides will engage in such anticipation, thus bringing a successful preemption by one side just at the moment when there seems no call whatsoever for alarm (this being by definition the only time for success). Also, the prospect of an attack based on dread of not attacking, mentioned in the third hypothesis, cannot be ruled out, if for no reason except fear that a spontaneous aggression might well be committed with no warning. Hence, to forestall the likelihood of an opponent's attacking without having carefully calculated all the consequences, it might perhaps seem wise to have nuclear weapons as a kind of strategic deterrent against an enemy's conventional attack that might succeed if tried. It is the judgment of the authors of this

study, as well as of others, that Israel now has nuclear weapons and that the Arabs will move to obtain them in the near future.[17] Some have gone so far as to argue that Israel should openly deploy nuclear devices as a stable deterrent in the Middle East—a veritable second-strike retaliatory capability.[18] What is often overlooked in this argument, however, is not that, one side having nuclear weapons, the other side might obtain them, but that, given the preemptive attack strategies now being developed, there will be a tendency to move nuclear weapons out of the deterrent realm into the attack strategy itself. In other words, where actual warfare exists, nuclear weapons will eventually be used.

(6) *Preemptive capabilities for Israel and the Arab states will become nuclear as well as conventional: with appropriate technology it is almost inevitable that nuclear weapons will find their way into battlefield and rear-echelon usage.* Most of the stand-off missiles, surface-to-surface missiles, and guided bombs are reckoned "dual capable" in American and Russian inventories, but are supposedly only conventional in the Middle East. Against targets in zones 2 and 3 nuclear weaponry would provide an even more convincing first-strike capability than conventional arms. The presence of a doomsday atomic weapon in the Middle East, therefore, would seem to be no more a stable deterrent than would preemptive capability itself—that is, it would provide no stable deterrence whatsoever. Indeed, with the presence of any nuclear arms in the Middle East, a signal that an attack is under way might signify to a fearful adversary that incoming weapons were nuclear as well as conventional. The prospect of nuclear attack would only increase fear, contributing to further unsettling of an already unsettled situation. If one were to have nuclear weapons in one's arsenal and would have to launch one's own preemption at the first warning that an opponent was attacking

[17] Among more recent analyses of Israel's nuclear capability see Pranger and Tahtinen, *Nuclear Threat in the Middle East*; William Beecher, "U.S. Believes Israel Has More than 12 Nuclear Weapons," *Boston Globe*, 31 July 1975; Arthur Kranish, "CIA: Israel Has 10-20 A-Weapons," *Washington Post*, 15 March 1976; David Binder, "CIA Says Israel Has 10-20 A-Bombs," *New York Times*, 16 March 1976. Israel has never officially confirmed reports that it has nuclear weapons, but instead has insisted it would not be the first nation in the Middle East to introduce such weapons into the region. This pledge has become somewhat clouded by the reports, never officially confirmed by the United States, that Soviet nuclear weapons had arrived in Egypt during the October 1973 war.

[18] See Robert W. Tucker, "Israel and the United States: From Dependence to Nuclear Weapons," *Commentary*, vol. 60, no. 5 (November 1975), pp. 29-43; and Steven J. Rosen, unpublished paper cited in Pranger and Tahtinen, *Nuclear Threat in the Middle East*, p. 5, n. 4.

with dual-capable weapons (even if the warning were mistaken), there would be a high probability that one would fire nuclear weapons at one's enemy.

Implications for United States Policy in the Middle East. A good deal of discussion has taken place about U.S. arms supply policies in the Middle East. Despite the controversy surrounding these policies, certain facts seem fairly well established. Fundamentally, Middle East actions by the superpowers are politically determined. The weapons supplied have increased in sophistication. The quantities of weapons provided have increased. And while weapons available may not be the fundamental cause of conflict in the region, their availability has certainly not brought peace.

The guidelines for U.S. arms policy in the Arab-Israeli conflict are basically political, as are the guidelines for the Soviet Union's policy. Neither Moscow nor Washington seems really capable of saying "no" to its clients, even though at times there is delay in supply and even bad feeling between client and superpower. Both the United States and Soviet Union are impelled by internal and external considerations to grant and sell large quantities of arms to the region of the world most prone to conflict, even though the risks of supplying these weapons may extend to the relations between the superpowers themselves.

Even though it is possible that our six hypotheses may be disputed and subject to reanalysis, they do provide at least a start toward developing guidelines for U.S. arms policy in the Arab-Israeli conflict—guidelines that do not depend only on the vicious-circle logic of competition with the Soviet Union and the realities of U.S. domestic politics. These hypotheses suggest a question of what kind of war the United States is prepared to condone in the Middle East. It is clear, in any case, that the present arms supply policies of the superpowers are tending toward preemptive capabilities for both the Arabs and Israelis, and that eventually both sides will be "dual-capable" with conventional and nuclear weapons.

Using the six hypotheses developed earlier as a point of departure, one may discern the outline of an evolving Arab-Israeli military relationship. (The outline may, of course, be altered at any given point by changes in U.S. military assistance doctrine in the Middle East.) Each element in this relationship raises a question for American policy.

(1) Because of necessarily inadequate warning, static defensive strategies are likely to be increasingly considered inappropriate. A

relevant question would therefore be, should U.S. policy concentrate more on distant warning systems or on providing newer weapons to negate such systems?

(2) Because of new weapons technologies, dynamic preemptive strategies are likely to be increasingly considered appropriate. A key question would therefore be, should the United States work toward agreements with the Soviet Union to withhold certain technologies from the Middle East or move to provide even more advanced technologies?

(3) Concepts of "balance" and "stability" in Arab-Israeli military relationships are likely to become progressively more obsolete. An important consideration would therefore be, should the United States improve understanding on both sides of the Middle East conflict about the other's capabilities or promote greater secrecy of efforts?

(4) The Middle East is likely to become more and more prone to instability, so that accidents and mistaken judgments will trigger outbreaks of warfare. Thus it would be essential to ask, should the United States establish more thoroughgoing crisis management machinery in the Middle East, with the Soviet Union and regional parties, or adopt a hands-off attitude?

(5) In possible local efforts to stabilize deterrence and develop a second-strike retaliatory force, nuclear weapons are likely to be seen as an ultimate security guarantee. Accordingly, should the United States actively encourage a nuclear ban in the Middle East, with Soviet help, or desist?

(6) With nuclear possibilities for retaliatory strategies realized, atomic weapons are likely to be incorporated in unstable preemptive strategies as well, compounding the already progressive instability. Therefore, it would be particularly relevant to ask, should the United States more closely monitor "dual-capable" arms in the Middle East, even if they are disarmed of nuclear devices, or should it adopt a more permissive approach?

Options for military assistance programs do not ordinarily develop out of thin air: they evolve from foreign and defense policy assumptions about what is desirable for the United States as well as for the countries being assisted. One set of assumptions might focus on preferred political and military outcomes—outcomes such as peace in the Middle East. Another set might deal with what uses the recipients will make of weapons supplied. Of course, the two sets of assumptions need not be antagonistic: a case can be made that both should be given some hearing. While focusing on the possible future uses of weapons in the Middle East, this study does not deny the importance of broader political-military goals pursued by the United States. On the other hand, it is suggested here that the broader goals themselves may be drastically and negatively affected by the ways weapons supplied are actually used in the event of future conflict. This same contradiction between ends and means applies to the Soviet Union; in its ostensible efforts to bring "justice"—for its Arab clients—to the Middle East, Moscow may be helping to pave the way for enormous destruction inimical to the region as a whole.[19]

2. LIKELY USES OF WEAPONS IN THE MIDDLE EAST

Dale R. Tahtinen

The breakdown of the Middle East into four possible zones of warfare (p. 1) provides a useful tool for viewing the types of weapons now or soon to be in the region. Since arms are ordinarily applied to those purposes for which they were developed, it is instructive to examine these purposes and on the basis of such an examination to suggest the likely uses to which Arab and Israeli weapons will be applied.

Before this examination is begun, some of the more significant aspects of warfare in the Middle East should be set out. Not surprisingly, the single most important goal for either side in any Arab-Israeli war is to gain control of the skies. Once control of the skies is achieved, the successful force can attack enemy targets with relative impunity in all four zones of battle. Thus quality of aircraft, air defenses, and electronic countermeasures are particularly crucial.

Aircraft. Combat aircraft are of special importance because they can be used in a number of ways to attack targets in all four battle zones. As Tables A-1, A-2, and A-15 indicate (see Appendix), there has been a great deal of emphasis placed upon the procuring of large quantities of highly sophisticated planes by both sides.

---

[19] Secretary Brezhnev avers that the Soviet Union is "searching for the solution of such a problem as ending the arms race in this region" [the Middle East]. "But it stands to reason that this must be tied in closely with a general settlement in the Middle East. To tackle the problem before such a settlement is reached would place the aggressor on a par with his victims." Report to the 25th CPSU Congress by Leonid Brezhnev, General Secretary of the CPSU Central Committee, 24 February 1976 (Soviet Embassy, Information Department, Press Release).

As of March 1976, the Israelis possess some of the finest American combat aircraft ever built, and when these are combined with highly effective American-supplied and Israeli-produced electronic countermeasure (ECM) equipment and with superior pilot skill, the result is to give Tel Aviv a decided edge in air superiority. These planes and the indigenously assembled Kfir, along with the F-15 slated for delivery in the near future, would allow Israel to make effective attacks on Arab targets in all zones, though there would certainly be some Israeli losses in penetrating the Russian-supplied air defenses on the Syrian and Egyptian fronts and the soon-to-be supplied U.S.-built Hawk air defense system to be installed in Jordan.

In any such attack against enemy targets in the four zones by either side, the role of electronic countermeasure will be of great importance. As of March 1976 Israel would probably prevail in this area but only after some losses. With a combination of jamming pods on the combat planes themselves, heliborne ECM platforms, and pilotless craft, Israel's forces possess significant offensive and defensive air capability. Of course, this equipment is also valuable when combined defensively with the U.S.-supplied Hawk system. In addition, the Israeli acquisition from the United States of at least four E-2C airborne early warning and fighter control planes (now on order) will significantly increase Israel's capability to operate in a hostile electronically cluttered environment.

Surface-to-Surface Missiles. While sophisticated aircraft will remain crucial in any future Arab-Israeli war, other relatively new weapons systems are already in—or are soon to be added to—the inventories of the warring states. Foremost among them are the different highly advanced surface-to-surface missiles (SSMs) and air-to-surface missiles (ASMs).

Unlike combat aircraft, which can be used effectively for strikes against targets in all four zones (whether used to attack or to provide air cover), the various types of SSMs are primarily directed against military targets. In a conventional mode, SSMs such as the Scud B and the longer range Jericho would most likely be used against targets in the last three zones and would probably be fired to strike those targets that are most heavily defended and therefore most likely to exact a heavy toll of attacking aircraft. Of course, as the distances to the targets increase, particularly in zones 3 and 4, these missiles are less and less likely to be used.

In a conventional mode such SSMs would most likely be used against air defense sites, or against large concentrations of men and materiel, and of course they could be fired against communications, transportation and economic centers. Scuds and Jerichos would be far more destructive per dollar invested (and therefore more efficiently used) with nuclear than with conventional warheads. In that mode they would undoubtedly be fired only against major targets some distance away, such as cities or particularly strong structures such as major dams.

Arab Frog-7s and Israeli Lance SSMs with their shorter ranges would be launched primarily against front-line battlefield targets in zone 1 and perhaps in zone 2. Once again, the range of the missiles would be a limiting factor. However, the Lance, presumably like the Scud, was developed for use with a nuclear warhead, and with such a warhead in that mode is designed for firing at large battlefield concentrations of men and materiel.[20]

Any discussion of missiles now or soon to be in the Middle East would not be complete without a mention of the Israeli interest in procuring the American Pershing missile and of the likely ramifications of such a procurement. The Pershing SSM with its range (much longer than the range of any missile currently in Israeli possession) would most likely (in the Middle East) be fired at major targets in zones 3 and 4.[21] However, it should be noted that the Israeli request for Pershings stirred up enough opposition in the United States to force at least temporarily shelving of the question whether Israel's defense forces should have an SSM of this sort even with assurances that nuclear warheads would not be employed.

In discussing the possibility of providing Pershings to Israel, it is also essential to discuss the Soviet-made Scaleboard, the SSM that Moscow would most likely send to at least one of the Arab states in an effort to balance any Israeli possession of the American counterpart missile. Being similar to the Pershing, the Scaleboard would most likely be fired by the Arabs against the same kind of targets that the Pershing would be fired against by the Israelis.

[20] See Pranger and Tahtinen, *Nuclear Threat in the Middle East*, pp. 30-31; and R. T. Pretty, ed., *Jane's Weapons Systems 1976* (New York: Franklin Watts, 1975), pp. 41-42, 44-45.

[21] For a more extensive discussion of military doctrine relating to the use of such missiles see Pranger and Tahtinen, *Nuclear Threat in the Middle East*. Also see "Israel Again Asks for Pershing Missile." Drew Middleton, author of this article, notes that Israel has reported some 350 Scuds now in Arab hands, mainly in Syria, but that U.S. defense department officials think this estimate "high." According to Middleton, the major fear of American officials is "that possession of missiles by both sides could turn a fifth Arab-Israeli conflict into one involving the destruction of cities on both sides." As analyzed in this study, the danger from missiles would be more complicated than this, though it would include zone 4 (civilian) targets.

The presence of these two missiles in the Middle East, along with the other shorter range SSMs, would guarantee that in a future conflict missile exchanges would take a serious toll (even if the missiles were conventionally armed) and would heighten the temptation for the two sides to use nuclear or other nonconventional (such as chemical or even biological) weapons.

Other significant kinds of SSMs include naval versions, which are of particular importance since any future Arab-Israeli war is likely to witness a significant amount of warfare conducted at sea. Foremost among these naval SSMs are the Styx (in Arab inventories) and Gabriel (Israeli-made) which have been the basic workhorse missiles in the past conflicts. Applying the four-battle-zone breakdown to the naval realm, we note that the Styx and Gabriel could be used in all four zones. Indeed, they could be used to fire at targets including not only vessels of war but also merchant ships. Attacks on merchant ships may be justified by a logic similar to that which justifies attacking the various infrastructures: the belligerents may deem it necessary to stop anyone's supplying the enemy's economy and to strike at rear-echelon targets from the sea when possible, as well as striking at coastal facilities and even at general population centers for psychological purposes.

The increased emphasis on naval battle preparations includes both sides' adding submarines to their inventories.[22] While it would seem short-sighted for either side to expand naval activities to include attacks on nonmilitary ships and civilian targets, naval preparations seem to indicate plans to do just that. In this regard reported comments from naval sources in Tel Aviv are especially sobering: they key on the necessity of protecting shipping lanes "from Sicily to the Straits of Bab-al-Mandab," and on Israel's perceived naval requirements. These perceived requirements have led to an offense-oriented defense that reportedly calls for seeking out "the enemy at his bases and his coasts." [23] Whether this means that in the future the Israelis will be attacking on the high seas or whether it means that they are fearful the Arabs will do so is unclear. However, one thing is clear—if warfare at sea begins to include transport vessels headed for the ports of either side, then the conflict will immediately affect a number

[22] The newest acquisitions have been by Israel when it added three German-designed submarines with a recently developed British missile system. There have also been reports that Libya will eventually be receiving some submarines from the Soviet Union.

[23] Herbert J. Coleman, "Israel Shifts toward Long-Range Fleet," *Aviation Week & Space Technology*, 10 March 1975, p. 57.

of nonbelligerent nations and could lead to a substantial expansion of the number of nations actively involved.

Air-to-Surface Missiles. Air superiority is of great importance in naval warfare, and nowhere is this fact better reflected than in the increasing emphasis being placed upon anti-ship air-to-surface missiles (ASMs) as well as on airborne anti-submarine weaponry (see, for example, Tables A-6 and A-8). For example, the Israeli-produced Gabriel can be launched from the air and the sixty-nautical-mile-range U.S.-produced Harpoon soon to be in Israeli inventories will be the air-launched version. Thus Israel is preparing an excellent capability for striking at all four zones of battle in any future war at sea.

Meanwhile on the Arab side there is no known airborne anti-ship capability that can compete with that of Israel. However, the Egyptians do possess a significant submarine force, and when this submarine force is combined with Egyptian and Syrian missile boats, it provides considerable naval strength.

The importance of air power is accentuated when the ground warfare role of air-to-surface weaponry is considered. The ASM and other "smart weapons" are of particular importance in zones 1, 2 and 3. Indeed, the description of the different kinds of ASMs in Table A-6 underlines the importance of air-launched missiles and the value of air superiority in gaining the best possible access position for firing at the chosen targets.

On Israel's side, an extensive variety and number of ASMs allows it to strike at targets in all four battle zones. In the first and second zones, Hobos and particularly the Shrike would be useful against surface-to-air missile (SAM) sites, while the Maverick and air-launched versions of the tube-launched, optically tracked, guided weapon (TOW) would be especially effective against Arab armor. Meanwhile, in the second zone, aircraft revetments, armor parks, and air defense systems are typical of the targets that could be hit with the same ASMs and with Walleyes. With the general exception of TOWs, these missiles could also be used against infrastructure targets in the third zone, especially inasmuch as major industrial, communication, and transportation targets would most likely be protected by some form of air defense. When the Condor is available for export to Israel, its long range and high degree of effectiveness will further widen the missile gap in Israel's favor.

In addition to ASMs, Israel also possesses laser-guided bombs that could be effectively used against targets in the rear-echelon areas and infrastructure. These precision-guided weapons could also be efficient in striking targets in zone 1.

Another important capability is the ability to use remotely controlled drones either to draw SAM fire before the arrival of aircraft with the sophisticated missilery they carry or to serve in an attack role (some of these pilotless craft can often carry 1,000 pounds of ordnance). Of course, drones can also be utilized in electronic countermeasure roles.

The Arabs are at a significant comparative disadvantage in ASM capability. Their primary reliance in this area is on the French Martel, which provides a good stand-off capability and could be used against air defenses in any of the battle zones. The Arab's disadvantage will continue until they can procure other missiles from western nations or from the Soviet Union, once that country develops such weapons fully and decides to supply them to its clients in the Middle East. It is, however, relatively certain that as long as the Israelis possess their increased margin of military superiority, the Arabs will make every effort to procure equipment similar to theirs, with the source of the equipment making little difference.

Technology Transfer. The existence of the different varieties of SSMs, of ASMs, and of the host of other weaponry listed in the tables obviously underlines the fact that the Middle East is a highly volatile region that could erupt into a war eventually encompassing the globe. Yet the sources of the danger are not limited to the armaments that might be imported into the area. Large amounts of technology permitting local production of advanced weapons have already been transferred into the region by the superpowers and the nations of western Europe, and the net result could well be to increase danger in an already explosive environment.

Even without Israeli nuclear capability and the Arab desire to offset it, the level of technological military knowledge in the area is awesome. For many, the fact that Israel has been able to develop an indigenously produced missile (Gabriel) and that the Egyptians made a significant effort to do likewise in the 1960s is highly alarming. Of even greater concern, however, are relatively new indications that Israel, along with a number of other countries, has the "technological infrastructure in electronics, airframes and jet engines to produce long-range cruise missiles."[24] If Tel Aviv proceeds to add weapons

[24] Stockholm International Peace Research Institute, *World Armaments and Disarmament*, SIPRI Yearbook 1975 (Cambridge, Mass.: MIT Press, 1975), p. 332.

such as these to its inventories, then (in the familiar cycle of the Arab-Israeli arms race) the Arabs will probably make every effort to procure equivalent weapons. The greater range for missiles means that more countries could well be included in future hostilities than have been included in the past.[25] Of course, unlike the case in which the United States could prevent Israel from acquiring the Pershing, there is no such prevention by an outside power when a country can produce its own sophisticated weapons. Perhaps the camel is already untethered (and with his nose poised just outside the tent), but the United States may still want to end the export of the kind of technology that has enabled Washington's client state to develop cruise-missile capability. Meanwhile, it may be astute for Israel not to develop such weapons or to widen its military superiority gap, hoping that under these circumstances Moscow will feel less pressured by the Arab states to provide them with such armaments or such technology. To develop such weapons—to widen the gap—would probably decrease Israel's long-range security and guarantee even more destructive wars in the future than have taken place in the past.

[25] An increase in the number of countries in which targets are hit by Israel in any future conflict would not be particularly surprising in light of warnings by Israeli leaders that other (undoubtedly the oil-rich) Arab states should realize that they too may be subject to Israeli punitive actions. In this regard, it may be noted that Tel Aviv's use of the C-130 as a paratroop carrier is consistent with the "stated Israeli goal of developing a strike force capable of hitting potential enemies as far from Israel as Libya or Kuwait." See "Israeli Paratroops Increase Mobility with C-130 Training," *Aviation Week & Space Technology*, 23 February 1976, p. 23.

APPENDIX

WEAPONS INVENTORIES IN THE MIDDLE EAST

Dale R. Tahtinen

The one phenomenon that has been consistently striking in the Arab-Israeli military balance is the sophistication and complexity of the armaments on each side. A highly destructive war was predictable in early 1973, and an even more potentially disastrous conflict may be on the horizon now. One need only look at the statistics on arms inventories to realize how portentous is the amount of military equipment available to the two sides in this always volatile conflict. Yet, as large as the numbers are, the preparation for war is even more awesome—given the relatively small size of the countries involved—when one considers the qualitative aspects of these armaments. The tables that constitute the bulk of this analysis are an effort to present Arab-Israeli armament comparisons in a concise, graphic manner.[1] Textual comments are brief, serving only to highlight particular points of which the reader should be cognizant.

Tables A-1 and A-2 give the quantitative and qualitative aspects of the arms race between the two sides for aircraft, while Tables A-3 through A-6 indicate the type and sophistication of missilery already or soon to be available to the Arabs and Israelis. Together, these two

---

This appendix is reprinted with minor revisions from Dale R. Tahtinen, *Arab-Israeli Military Status In 1976* (Washington, D. C.: American Enterprise Institute, 1976) which the present study replaces.

[1] Unless otherwise noted, the data in the subsequent tables rely upon the following sources: *Jane's All the World's Aircraft* for the years 1967-1968 through 1975-1976, along with *Jane's Fighting Ships* for the same years and *Jane's Weapons Systems* for 1971-1972, 1972-1973, 1974-1975, and 1976; *The Military Balance* 1967-1968 through 1975-1976 (London: Institute for Strategic Studies), *Aviation Week & Space Technology* for 1971 through 1975, and *The International Defense Review.*

sets of tables show the impressive buildup of destructive air power in this always dangerous area. One major element not reflected is Israel's qualitative advantage in terms of pilots and ground maintenance crews that keep a maximum number of planes available for combat duty.[2]

The importance of air superiority should be kept in mind in an examination of Tables A-7 and A-8 because, once control of the skies is established, helicopters can be of major value in providing the mobility essential to assist in seizing and maintaining initiative in combat. The multi-purpose role of helicopters extends from use as an attacking gunship to the transport of men and materiel to use in an electronic countermeasures role.

Table A-9 is of critical quantitative importance since one must have the troops in order to make an army function. It is unfortunate, however, that a characteristics table could not be built to reflect the qualitative factors which are crucial to the effective functioning of any armed forces. Suffice it to say that although the Arab officers and soldiers showed a marked improvement during the October war, the skill of Israeli manpower, whether on the battlefield or in reserve, generally continues unequaled in the region.

Tables A-10 through A-13 show armor and corresponding anti-armor equipment, which remain vital factors in any future hostilities. Indeed, the effectiveness of anti-armor equipment in destroying large amounts of tanks and armored personnel carriers in 1973 certainly did not lessen the demand for these old stand-by weapons, but merely accelerated the rush to procure improved versions and also to stockpile (and absorb into tactics) ever newer and more sophisticated anti-armor equipment.

A growing emphasis by both Arabs and Israelis on building navies with considerable range, presumably to protect shipping and to conduct offensive warfare at sea, is reflected in Table A-14. The advent of additional missile boats and submarines, combined with substantial airborne naval warfare capabilities (particularly on the Israeli side), indicates that any future conflict will undoubtedly see more action at sea than have the Arab-Israeli conflicts of the past.

---

[2] For an extensive treatment of the various qualitative factors and the different aspects of the arms race such as the electronic countermeasures competition, as well as a discussion of nuclear capabilities, see Dale R. Tahtinen, *The Arab-Israeli Military Balance since October 1973* (Washington, D. C.: American Enterprise Institute, 1974); Dale R. Tahtinen, *The Arab-Israeli Military Balance Today* (Washington, D. C.: American Enterprise Institute, 1973), pp. 32-36; and Pranger and Tahtinen, *Nuclear Threat in the Middle East.*

Table A-15 highlights some of the newer weapons likely to be introduced into the region in the absence of either a permanent peace settlement or some type of arms control arrangements.

The last table, drawn from a 1975 analysis entitled *Nuclear Threat in the Middle East,* focuses on the nuclear-capable delivery systems now deployed in the area. So much serious discussion of this issue has taken place since the appearance of this study that it is important to end these comments with a reminder that weapons sophistication of the kind now present in this region—when and if used—can lead to escalation across the threshold between conventional and nuclear forms of warfare.

Table A-1

COMBAT AIRCRAFT INVENTORIES IN THE MIDDLE EAST, 1967/68–1975/76

Country/Aircraft Type	1967-68[a]	1968-69	1969-70	1970-71	1971-72	1972-73	1973-74	1974-75	1975-76
EGYPT									
MiG-23 fighter	—	—	—	—	—	—	—	—	Some[b]
Mirage V ground attack	—	—	—	—	—	—	—	38	—
MiG-21 interceptor	100	110	100	150	200	220	210	200	250
MiG-19 fighter	45	80	—	—	—	—	—	—	—
Su-7 fighter-bomber	—	40	90	105	110	120	80	100	80
MiG-15 fighter					—	—	—	—	—
MiG-17 fighter-bomber	60	120	120	165	200	200	100	100	125
Il-28 light bomber	20	40	30	28	25	10	5	5	5
Tu-16 medium bomber	—	10	12	15	18	18	25	25	25
Total	225	400	352	463	553	568	420	468	485[b]
SYRIA									
MiG-23 fighter	—	—	—	—	—	—	—	Some	45
MiG-21 interceptor	—	60	55	90	100	140	200	200	250
Su-7 fighter-bomber	—	20	20	40	30	30	30	30	45
MiG-17 day fighter/ ground attack	—	—	—	—	—	80	80	60	50
MiG-15 fighter	25	70	70	80	80	—	—	—	—
Il-28 light bomber	—	—	—	—	—	—	—	Some	Some
Total	25	150	145	210	210	250	310	290[c]	390[c]

JORDAN									
F-5A fighter	—	—	—	—	—	—	—	—	24
Hunter ground attack	—	12	11	20	18	35	32	32	—
F-104A fighter-bomber	—	—	—	18	15	15	20	18	18
F-86	—	4	—	—	—	—	—	—	—
Total	—	16	11	38	33	50	52	50	42
IRAQ									
MiG-23 fighter	—	—	—	—	—	—	—	—	30
MiG-21 interceptor	50	60	60	60	85	80	90	100	100
Su-7 fighter-bomber	—	20	20	50	48	48	60	60	60
MiG-19 fighter	34	45	45	45	—	—	30	—	—
MiG-17 fighter-bomber					15	20	—	30	30
Tu-16 medium bomber	6	8	8	8	9	9	8	8	7
Il-28 light bomber	10	10	10	10	12	—	—	—	—
Hunter ground attack	50	50	50	36	35	32	36	20	20
T-52 light strike	20	20	20	20	16	—	—	—	—
Total	170	213	213	229	220	189	224	218	247
LIBYA									
Mirage V and III/B/C/E ground attack and interceptor	—	—	—	—	—	72	110	62[d]	92
F-5A fighter	—	—	—	7	7	10	9	—	—
Total	—	—	—	7	7	82	119	62	92

Table A-1 (continued)

Country/Aircraft Type	1967-68[a]	1968-69	1969-70	1970-71	1971-72	1972-73	1973-74	1974-75	1975-76
ALGERIA									
MiG-21 interceptor	— (MiG-21 and MiG-15, -17 combined)	140 (MiG-21 and MiG-15, -17 combined)	140 (MiG-21 and MiG-15, -17 combined)	140 (MiG-21 and MiG-15, -17 combined)	30	30	35	35	35
MiG-15, -17 fighter-bomber					60	95	95	95	80
Su-7 fighter-bomber	—	—	—	—	—	—	20	20	20
Il-28 light jet bomber	—	30	30	30	24	30	30	30	25
Magister armed trainer	—	—	—	—	28	26	26	26	26
Total	—	170	170	170	142	181	206	206	186
SAUDI ARABIA									
F-5E, -5B fighter	—	—	—	—	—	—	—	34	30
BAC-167 ground attack	—	—	—	24	20	20	—	21	30
Lightning fighter	4	24	28	35	20	35	—	35	35
Hunter ground attack	4	4	4	—	—	—	—	—	—
F-86 fighter	12	11	11	16	15	15	—	—	—
Total	20	39	43	75	55	70	—	90	95
ISRAEL									
RF-4E Reconnaissance	—	—	—	—	—	—	—	6	6
F-4 fighter-bomber/interceptor	—	—	—	36	75	120	127	150	200
A-4 fighter-bomber	—	48	48	67	72	125	162	180	200

Mirage III fighter-bomber/interceptor	65	65	65	60	60	50	35	25	110[g]
Kfir fighter	—	—	—	—	—	—	25	30	—
Vautour light bomber	15	15	15	12	10	10	12	10[e]	X
Mystere IVA fighter-bomber	25	35	35	30	27	27	23	23[f]	X
Oragan fighter-bomber	50	45	35	30	30	30	30	30[e]	X
Super Mystere interceptor	25	15	12	10	9	9	18	12	X
Total	180	223	210	245	283	371	432	466	516

[a] All data in this table and those following are generally from 30 June to 30 June.

[b] 48 MiG-23 are reportedly scheduled to be delivered. However, since the Sinai accord there has been some question whether the Soviet Union will provide the aircraft.

[c] Some aircraft are believed to be in storage. The MiG-25s, being flown in the Middle East, which are reconnaissance versions, appear to remain under Soviet control.

[d] Some Mirage may be in storage and 38 Mirage Vs are in Egypt.

[e] In storage.

[f] In reserve.

[g] Mirage IIIs and Kfirs combined. There are more than 110 Kfirs in the Israeli air force as of January 1976 with the number increasing to a target requirement of 200. The number of Mirage IIIs has been decreasing, the latest figure being 25 in 1974-75. Israel has been engaging, however, in project "Salvo" designed to prolong the usefulness of the Mirage III by replacing the existing Atar engines with the General Electric J-79 engine which is also used in the Kfir.

X = Unspecified number in reserve.

Table A-2

TECHNICAL CHARACTERISTICS OF MAJOR ADVANCED AIRCRAFT IN THE MIDDLE EAST

Aircraft Type	Ordnance Capability[a] (tons)	Approximate Combat Radius (miles)	Avionics, Weaponry and Operational Role
MiG-25	Unknown	700	Little is known about the avionics, but the aircraft was probably designed to intercept fast-strike aircraft, possibly with "snap-down" missiles to deal with low-flying raiders. It presumably carries air-to-air guided weapons. The version in Middle East is apparently under Soviet control and appears to be the reconnaissance type.
MiG-21	.6	350	The MiG-21 is fitted with search-and-track plus warning radar and Atoll air-to-air missiles, with a probable infrared guidance system similar to that of the American-made Sidewinder 1A.
MiG-23	4	520nm	A tactical air superiority fighter designed to intercept fast-strike aircraft. Little is known for certain about its avionics, but the MiG-23 is believed to have several of the following features: radar, ECM equipment, and possibly "snap-down" missiles for use against low-flying attack aircraft. It can carry a variety of armament.
Su-7	2.2	200-300	Subsequent to 1961, the Su-7 became the standard tactical fighter-bomber of the Soviet air force. Little information is available on its avionics.
Mirage V	4.4	400-805	The Mirage V can carry one Matra R-530 all-weather air-to-air missile with radar or infrared homing heads, an air-to-surface missile, and two Sidewinder air-to-air missiles. Avionics include air-to-air interception radar, with an additional mode for control from the ground and a sighting system giving air-to-air capabilities for cannons and missiles and an air-to-ground capability for dive bombing.
F-4E	8	900-1000	The F-4E uses highly sophisticated electronic countermeasures equipment, computers, and radar in its role as a long-range all-weather attack fighter. It is considered to be the best operational American missile-armed aircraft.

A-4E	5	400-800	The A-4E fighter-bomber is equipped with an angle-of-attack indicator, terrain-clearance radar, and a variety of optional sophisticated weapons such as air-to-air and air-to-surface rockets (Sidewinders, infrared Bullpups, air-to-surface missiles and torpedoes).
Mirage IIIC	1	560-745	The characteristics of the Mirage IIIC are the same as for the Mirage V, except that the planes supplied to Israel are said to have a different electronic configuration. Israel has developed an infrared homing air-to-air missile with a "see-and-shoot" capability and has fitted it to its Mirage fighters.
Kfir	4	320-780	A dual-purpose fighter-bomber with advanced avionics. Based on the Mirage V, Kfir is designed for low-altitude, high-speed performance—both for attack and dogfight missions. Kfir can carry a variety of weapons, including a 30mm cannon, four Shafrir missiles, Maverick missiles, Hobo Mk84 guided bombs, Shrike missiles, Luz missiles (under development in Israel), as well as conventional bombs and rocket pods.
F-5E	3.5	165-760nm	A light tactical fighter whose emphasis is on maneuverability rather than high speed. F-5E carries radar for air-to-air search and range tracking, two AIM-9 Sidewinder missiles, two 20mm cannons, and up to 7,000 lbs. of mixed ordnance including bombs, napalm, and rockets. Maximum speed: Mach 1.6. Combat radius: 165 nm (with maximum ordnance +5 minutes combat), 760 nm (with two Sidewinder missiles, maximum fuel +5 minutes combat).

Table A-3

MAJOR SURFACE-TO-AIR MISSILES IN THE MIDDLE EAST

Type	Characteristics	Effective Ceiling (feet)
SA-2	The Soviet-made SA-2 is a two-stage missile with automatic radio command. Target aircraft are tracked by radar which feeds signals to a computer, from which radio signals to the missile are generated. Those captured by the Israelis in 1967 had warheads detonated by contact or proximity fuses. American pilots in Vietnam reported that the SA-2s there lacked the maneuverability needed to intercept high-speed aircraft performing evasive tight turns.	60,000
SA-3	The SA-3 is a Russian-produced, two-stage, compact, mobile land-based missile used for short-range defense against aircraft beginning at low altitudes. It is similar to the American made Hawk used by Israel.	40,000
SA-6	This Soviet missile is considerably larger than the American-built Hawk but is assumed to have the same basic low-altitude defense role. Each unit consists of a tracked vehicle with three solid propellant missiles.	55,000
SA-7	The SA-7, an optically guided missile with shoulder-firing capability, is used against low-flying aircraft. It carries an infrared warhead. The Arabs successfully used this missile to cripple a number of Israeli planes in the October war.	3,000
SA-9	Probably designed for low-flying interceptors. The SA-9 is believed to have an active radar homing system. Little else is known about this Russian missile now thought to be in Arab hands.	Unknown
Hawk (MIM-23A)	This is an American-made guided weapon with a solid-propellant motor and a continuous-wave, semi-active radar-homing system. It is said to be effective against aircraft flying at normal combat altitudes down to tree-top level. Israel uses the Hawk and Jordan will be receiving it soon. Both will rely heavily on the new improved Hawk which is considered to be even more effective.	Over 38,000
Redeye	An optically guided American-made SAM with shoulder-firing capability. Designed for use against low-flying aircraft, Redeye has an infrared homing system which carries a high explosive warhead. It is now in Israeli inventories and soon will be in those of Jordan.	Probably over 6,000

Table A-3 (continued)

Type	Characteristics	Effective Ceiling (feet)
Rapier	This is a highly mobile, British-produced guided-missile system for use against fast, low-flying aircraft. It is commanded to follow an optically established sightline to the target. The Rapier's direct-hitting system has been repeatedly successful in a number of trials. Egypt is likely to begin procuring sizable amounts shortly.	Believed to be over 15,000
Chaparral	This mobile, American-produced missile system is of an infrared heat-seeking variety developed for low-altitude requirements. Chaparral has an amphibious capability when mounted on an M-730 self-propelled vehicle with a "swim kit" and can be used in conjunction with the Vulcan air-defense system. Israel has added this to its Hawk-Vulcan air defenses.	Low altitude

Table A-4

MAJOR SURFACE-TO-SURFACE MISSILES IN THE MIDDLE EAST

Type	Characteristics
Frog-7	An unguided, spin-stabilized Soviet tactical missile, fired from a modern wheeled erector-launcher and possessing either a high-explosive or nuclear warhead. Frog-7s were used during the 1973 Arab-Israeli war and gave rise to speculation that a radio-command guided version exists and that this may have been the model used by the Arabs in the war. Range: over 35 miles.
SCUD B	A Russian rocket guided by movable tail fins, and believed to have a simplified inertial guidance system. SCUD B can carry either a high-explosive or nuclear warhead. It is believed to be in Arab inventories. Range: some 175 miles.
Samlet	A Russian jet-powered cruise missile used by the Arabs. Believed to possess a semi-active homing seeker, Samlet probably has a command guidance system. It is used for coastal defense. Range: about 125 miles.
Jericho	A 280-300-mile-range battlefield support missile believed to be in production in Israel. No other specifications are currently available.
Ze'ev (Wolf)	Not much is known about this Israeli short-range missile, which comes in two versions: one has a 170kg warhead and a range of 1,000 meters, the other a 70kg warhead and a range of 4,500 meters. Used during the Arab-Israeli war of 1973, it is not believed to be very accurate, but is considered effective against large concentrations of men and vehicles.
Lance	This highly mobile American missile is considered to be very accurate. It uses a simplified inertial guidance system and was initially developed for use as a tactical nuclear delivery system, although a nonnuclear warhead is presently being produced. The first installment of 109 was scheduled to arrive in Israel by the end of 1975, and Israeli Lance crews have been in training since early last year.
Gabriel	There are two versions of this Israeli-developed anti-ship missile which performed well during the October war. The missile uses a semi-active radar homing system and it is likely that an alternative form of homing is available for application in an ECM environment. Propulsion is by a two-stage rocket motor. The range is up to 14 nm on the first generation version. The second, called Gabriel II, is said to be similar but probably with a larger motor that reportedly nearly doubles the range.
Styx	This Soviet-produced anti-ship missile has been improved over its lengthy time of service. It is thought that Styx can be guided in the cruise phase either by auto pilot or radio command, and in the terminal phase—either by continuing command, active radar or infrared homing. Styx is used on the Arab missile boats. Estimated range: up to 20 nm.

Table A-5

MAJOR AIR-TO-AIR WEAPONS ALREADY OR SOON TO BE IN THE MIDDLE EAST

Type	Characteristics
Atoll	An infrared guided Soviet-made missile which carries a conventional high-explosive warhead. Atoll closely resembles the U.S.-made Sidewinder and is carried on the MiG-21.
Matra R.530	An all-weather, French-produced missile, with alternative semi-active radar or infrared homing guidance heads. The infrared homing version can successfully intercept aircraft by relying on heat from airframe hotspots when attacking the front or from engines and exhausts when attacking from the rear. Use of the proportional navigation system in the semi-active radar homing version allows for a wide firing range and good hit probability against maneuvering targets. All three versions listed probably either are, or soon will be, on Arab Mirages. Some are also in Israeli hands.
Matra Super 530	A revision of the R.530 missile designed to meet the high speed and altitude requirements of the next generation of interceptors by doubling its predecessor's capabilities in range and acquisition distance. Equipped with an electromagnetic homing head, the new missile has an all-weather capability.
Matra R.550 magic	A new, air superiority missile designed for close combat. The R.550 Magic has an infrared homing guidance system and a canard arrangement of the forward fins for twist and steer control.
Sidewinder AIM-9C/D/E/G	A short/medium range American missile. The 9C has a semi-active radar homing guidance system; the 9D has an infrared guidance system; the 9E has an improved infrared homing head and other modifications designed to increase low-altitude performance; and the 9G incorporates an Expanded Acquisition Mode which increases lead acquisition capability. The radar homing system of the 9C can use either the target illumination radar of the launch aircraft or the radiation of the ECM signals of the target aircraft. Some versions are already in Israeli inventories.
Sparrow III AIM-7	One of the most successful and widely used U.S.-made weapons, the Sparrow is guided by semi-active CW radar. One version is designed for greater maneuverability in close combat. The newer 7F possesses solid state electronics for its semi-active radar and a larger motor for increased range. Some versions are already in Israeli inventories.
Shafrir	This Israeli-produced missile is of the infrared homing variety and has reportedly achieved a high kill ratio in combat. One of its practical features is that opposition pilots seldom have an opportunity to escape. Electronics are solid state and firing is on a "see and shoot" basis.

Table A-6

MAJOR AIR-TO-SURFACE WEAPONS ALREADY OR SOON TO BE IN THE MIDDLE EAST

Type	Characteristics
Kelt (AS-5)	Kelt, which is carried by the TU-16 bomber, uses a rocket propulsion motor and has a maximum range of some 200 miles.
Condor (AGM-53A)	A medium-range American cruise missile whose primary roles are stand-off missions against heavily defended, high-value surfaced targets. Developed to replace Bullpup missiles, Condor has a TV remote guidance and control system, carries a high-explosive warhead, and possesses the capacity to carry a nuclear warhead as well. A new version is being developed which will permit a night all-weather capability. This missile is intended to operate successfully in hostile electromagnetic environments and will have the data link system of the Walleye II ASM with dual mode radar and electro-optical seeker. Condor reportedly may be sent to Israel as part of the Sinai aid package. Range: 35 to 50 miles, probably to be increased.
Walleye (AGM-62A)	An unpowered American TV-guided glide bomb that comes in several versions. The newer one possesses a data link and has an extended range. Walleye is used against semi-hard targets such as bridges, air-base facilities and ships. Israel has received some Walleyes.
Maverick (AGM-65A)	A small, TV-guided tactical missile designed for use against small targets such as armored vehicles, parked aircraft, and field fortifications. Maverick carries a 59kg high-explosive warhead which is conical-shaped for high penetration. A laser-guided version is now being developed which will provide a capability for night attack against air defenses. Maverick was used effectively by Israel in October 1973.
Martel	An Anglo-French stand-off missile that comes in two versions offering alternative terminal guidance systems. One (AS.37) is the anti-radar model with a passive radar homing system. The other (AJ.168) has a visual guidance system using a TV camera and a data link which relays video and command guidance signals between the missile and parent aircraft. Guidance for the initial cruise phase of both models is provided by an auto-pilot system. Both versions carry a proximity fused high-explosive warhead. The Martel will soon be available in significant numbers for Arab Mirages. Range: over 35 miles.
Harpoon (AGM-84A)	An American all-weather anti-ship missile capable of being launched from both ships and aircraft. Harpoon's mid-course guidance system employs a digital computer and a strap-down sensor/guidance subsystem—a technique comparable to inertial guidance; its terminal phase is guided by active radar homing. This missile can carry a 500 lb. high-explosive warhead. Harpoon has been approved for delivery to Israel. Range: some 60 nm.

Table A-6 (continued)

Type	Characteristics
Shrike (AGM-45A)	An anti-radiation ASM designed for use against ground defensive radar installations. Shrike's radar sensor head detects the direction of radar radiation and commands the guidance system to home in on the radar, making this missile effective against enemy early warning, ground control intercept, and SAM guidance radars with their different frequencies. This American missile is in Israeli hands. Range: 7-10 miles.
Hobo	An electro-optically guided smart bomb used primarily against ground-based air-defense installations. Using a TV monitor and data-link guidance system, Hobos are guided by a missile operator in the parent aircraft who, depending upon the model, either aims the bomb or controls the bomb's glide through manipulation of its wings until target acquisition, when the terminal homing system takes over. Other models have their own independent navigation system which employs a radar homing technique. This and similar laser-guided weapons will likely be made available in large numbers to Israel as part of the Sinai aid package. Hobo comes in 2,000 and 3,000 lb. sizes.

Table A-7

HELICOPTER INVENTORIES IN THE MIDDLE EAST, 1967/68–1975/76

Country/ Aircraft Type	1967-68	1973	1974-75	1975-76
EGYPT				
Commando	—	—	—	24
Mi-4 and Mi-6	30	—	—	—
Mi-1, 4, 6 and 8	—	190	200	110
Sea King	—	—	30	4
Total	30	190	230	138[d]
SYRIA				
Mi-2	—	—	4	4
Mi-1	7	?[a]	—	—
Mi-4	3	?[a]	8	8
Mi-8	—	?[a]	39	39
Ka-25	—	—	10	9
Total	10	50	61	60
JORDAN				
Alouette III	4	6	6	10
Whirlwind	4	3	3	3
Total	8	9	9	13
IRAQ				
Mi-6	—	—	16	16
Mi-1	—	—	—	—
Mi-4	9	35	35	35
Mi-8	—	29	30	30
Alouette III	—	5	20	20
Wessex	11	—	—	—
Total	20	69	101	101
LIBYA[b]				
AB-47	—	—	—	6
AB-206	—	2	5	2
OH-13	—	3	7	7
Alouette III	—	10	10	10
Super Frelon	—	9	9	9
Total	—	24	31	34

Table A-7 (continued)

Country/ Aircraft Type	1967-68	1973	1974-75	1975-76
ALGERIA				
Mi-8	—	—	—	5
Mi-6	—	—	—	4
Mi-1	—	4	4	—
Mi-4	50[c]	42	42	42
Hughes 269A	—	6	6	6
SA-330	—	5	5	5
Total	50	57	57	62
SAUDI ARABIA				
Alouette III	2	1	—	—
AB-204	—	1	—	—
AB-205	—	8	10	10
AB-206	—	20	20	20
Total	2	30	30	30
ISRAEL				
S-65	—	—	—	20
CH-53G	—	12	18	18
UH-1D Iroquois	—	25	25	25
AB-205	—	20	20	20
Super Frelon	40 (Super Frelon, Alouette II, S-58 combined)	12	9	9
Alouette II		—	5	5
S-58		—	—	—
Total	40	69	77	97[e,f]

[a] Number is unknown.

[b] In the summer of 1970 Libya had some helicopters, including three Alouette IIs.

[c] The first information available was in 1969-70.

[d] In 1976, Egypt is scheduled to receive forty-two SA-342 French-English helicopters which are capable of carrying a variety of weaponry.

[e] Israel has requested, and reportedly may receive, a significant number of AH-1 HueyCobra gunships. This sophisticated American-made helicopter can carry a variety of weapons and perform many different combat roles.

[f] It should also be noted that Israel has been using its C-130s, of which it has 15, for paratroop drops. This development greatly increases Tel Aviv's potential radius of action.

Table A-8

CHARACTERISTICS OF MAJOR HELICOPTERS IN THE MIDDLE EAST

Type	Characteristics
UH-1D Iroquois	The Iroquois can carry a pilot plus twelve troops and has a maximum takeoff and landing weight of 9,500 lbs.
AB-205A	The AB-205A is a multi-purpose helicopter similar to the UH-1D. It can carry up to fourteen passengers plus pilot and has a maximum takeoff weight of 10,500 lbs.
AB-206A	The AB-206A carries four passengers in addition to the pilot and has a cruising speed of 130 mph with a weight of some 3,000 lbs.
AB-212	The standard accommodation of the AB-212 is fourteen passengers and pilot. The aircraft can carry up to 11,200 lbs. and its cruising speed is 127 mph.
CH-53	The CH-53 accommodates two pilots plus thirty-eight troops and has a maximum takeoff weight of about 42,000 lbs.
Mi-8	The Mi-8 carries two pilots and twenty-eight to thirty-two passengers. This transport helicopter has a cruising speed of 112 to 140 mph.
Mi-6	The Mi-6 is a transport helicopter which carries a crew of five and up to sixty-five passengers. Maximum cruising speed is 155 mph.
Super Frelon	The Super Frelon carries a crew of two plus twenty-seven to thirty troops and has a maximum takeoff weight of 28,600 lbs.
Westland Sea King	An advanced ASW helicopter with a capability for a number of other roles, such as search and rescue, tactical troop transport casualty evacuation, cargo carrying, and long-range self-ferry. Sea King can carry four Mk44 homing torpedoes or four Mk11 depth charges. It can also carry Martel air-to-surface missiles. Maximum load capacity: 8,000 lbs. Speed: 143 mph. Ferry/transit range: 520 nm with standard fuel; 750 nm with auxiliary fuel.
Commando	A tactical helicopter whose capabilities in payload and range performance represent an optimization of those of the Sea King. Commando's primary roles include ASW, tactical troop transport, logistic support, cargo transport, and casualty evacuation. Its secondary roles include air-to-surface strike and search-and-rescue. The Mk1 version carries twenty-five troops and a wide range of guns, missiles, et cetera. Cruising speed: 129 mph. Range with maximum payload: 305 nm; ferry range with auxiliary fuel: 814 nm.
AH-1G HueyCobra	An armed helicopter equipped with 7.62mm machine gun, 49mm grenade launcher, M-61 20mm Vulcan gun, and rockets, minigun pods, or TOW anti-tank wire-guided missiles. The AH-1Q TOW/HueyCobra model is designed to carry eight TOW missiles. Another model, the AH-1J Sea Cobra, is equipped with Marine avionics. Typical range: 425 miles. Speed: 219 mph.

Table A-8 (continued)

Type	Characteristics
KA-25	An ASW and general-purpose helicopter which has replaced the Mi-4 in the Soviet Navy. The ASW version has a search radar installation, a towed magnetic anomaly detector, and an electro-optical sensor. The KA-25 carries internally stored ASW torpedoes and can deliver nuclear depth charges. Maximum payload: 4,400 lbs. Normal cruising speed: 120 mph. Range: 217 nm typically, and 351 nm with maximum fuel.
AB-47 Agusta-Bell Model 47	A three-seat general utility helicopter. An ASW version (AB-47J-3) has been developed that carries one Mk44 torpedo. Speed: 105 mph. Maximum range: 182 nm.

Table A-9

ARMY STRENGTHS IN THE MIDDLE EAST, 1975/76

Country	Number
Egypt	275,000
Syria	150,000
Jordan	75,000
Iraq	120,000
Algeria	55,000
Libya	25,000
Saudi Arabia	40,000
Israel	400,000[a]

[a] At full mobilization, which takes 72 hours. The Arabs already are essentially at full mobilization. Their 1,000,000 soldiers listed as reserves would likely be of little additional assistance during hostilities given their training, equipment, and basic allotted role.

Table A-10

NUMBER OF TANKS IN THE MIDDLE EAST, 1968–1975/76

Country/ Vehicle Type	1968	1973	1974-75	1975-76
EGYPT				
JS-3	20	30	X	25
T-62	—	100	X	820
T-54/55	250	1,650	X	1,100
T-34	70	100	—	—
Pt-76	—	75	X	30
Centurion Mark 3	30	—	—	—
Total	370	1,955	2,000	1,975
SYRIA				
T-62	—	X	500	700+
JS-3	—	30	—	—
T-54/55	—	900	1,000	1,300
T-34	200	240	100	100
T-54	150	—	—	—
Pt-76	—	100	70	70
Old German tanks	50	—	—	—
Total	400	1,270+	1,670	2,170+
JORDAN				
M-60	—	} 200	240	240
M-47 and M-48	—			
Centurion	50	220	250	200
M-48	50	—	—	—
Total	100	420	490	440
ISRAEL				
Other medium tanks	—	—	—	350
Pt-76	—	—	X	65
T-62	—	—	X	150
M-60	—	150	X	450
M-48	225	400	X	400
Ben Gurion	250	250	—	—
Centurion	—	600	X	900
Sherman, Isherman, and Super Sherman	—	200	X	200
Super Sherman	175	—	—	—

Table A-10 (continued)

Country/ Vehicle Type	1968	1973	1974-75	1975-76
AMX-13	140	—	—	—
TI-67	—	100	X	—
T-54	200	—	X[a]	400
Total	990	1,700	1,900+	2,915

[a] Includes T-54/55.
X = Included, but exact number unspecified.

Table A-11

NUMBER OF ARMORED PERSONNEL CARRIERS IN THE MIDDLE EAST, 1973/74–1975/76

Country	Vehicle Type	1973-74	1974-75	1975-76
Egypt	BTR-40, 50P, 60, 152 and OT-64	2,000	2,000	2,500
Syria	BTR 50/60 and 152	1,000	1,400	1,100
Jordan	M-113	280	280	320
	Saracen	120	120	120
Israel	M-2, M-3, M-113 and salvaged captured APCs, BDRM, BTR-40, -50P/OT-62, -60P, -152	Over 3,000	2,500	3,300

Table A-12

CHARACTERISTICS OF MAJOR TANKS IN THE MIDDLE EAST

Type	Characteristics
T-62	This is believed to be standard equipment in the Soviet armored forces. It has a 115-mm. smooth-bore gun and a top speed of about 30 mph. It can cross water up to about eighteen feet in depth and has night-vision equipment.
T-54/55	Some are still in service with the Soviet armored forces. It is equipped with a 100-mm. gun and has a road speed of about 30 mph. It can cross water up to about eighteen feet in depth and has night-vision equipment.
Pt-76	The Pt-76 is a light amphibious tank used as the main reconnaissance vehicle of the Soviet army. It is capable of operating in a fast-flowing river. It is considered mobile but has limitations as a fighting vehicle. Armor protection is less than that of other light tanks. This vehicle has a 76-mm. low-velocity gun and a road speed of about 25 mph. It apparently has no night-vision equipment.
M-60	The M-60 is currently a main battle tank of the U.S. army. It carries a 105-mm. high-velocity gun, has a top speed of about 30 mph, can cross water up to about thirteen feet in depth, and is equipped with night-vision equipment. Its cross-country mobility is said to be inferior to that of more modern European tanks.
M-48	The M-48 is the main tank armament of the U.S. Marine Corps. It has a 90-mm. M-41 gun and a road speed of about 30 mph. Night-vision equipment can be fitted.
Centurion (Mark 13)	After twenty years of use, this tank is being phased out of the British army. The most recent version has a 105-mm. gun and a maximum speed of about 20 mph. It is fitted with night-vision equipment and can be fitted with water-crossing equipment.
Ben Gurion	This tank is similar to the Centurion, but the Israelis have replaced the original gun with a French 105-mm. gun.
Super Sherman	This is the old American Sherman tank modernized by Israel with the addition of French medium-velocity tank guns and new diesel engines. The majority of guns fitted are 105-mm., some are 75-mm. Road speed is less than 30 mph. Comparatively thin armor makes the Super Sherman vulnerable to the 100-mm. guns of the T-54/55.
Isherman	This is a rebuilt Sherman tank with a French 105-mm. gun and new engines for increased speed.[a] It is probably very similar to the Super Sherman.
Tl-67	Equipped with a 105-mm. gun, this is the Israeli conversion of captured T-54/55 tanks.
Tsabar (or Sabra)	Unconfirmed reports indicate that Israel has this main battle tank in development. It is said to be fitted with a British 105-mm. gun and designed specifically for desert conditions. It may have already entered service with Israeli armored forces.

[a] Christopher Foss, *Armoured Fighting Vehicles of the World* (New York: Scribner, 1971), p. 68.

Table A-13

MAJOR GROUND-LAUNCHED ANTI-TANK WEAPONS IN THE MIDDLE EAST

Type	Characteristics
Snapper (AT-1)	A Russian-made wire-guided, anti-tank missile launched from a guide rail mounted on a BDRM armored amphibious vehicle. Possessing vibrating trailing edge spoilers, Snapper can be fired and guided by an operator up to some 150 feet away from the launcher. Like Swatter and Sagger, the Snapper is in Arab inventories. Range: 500 to 2,300 meters.
Swatter (AT-2)	Similar to the Snapper. Swatter is a wire-guided missile which is believed to operate like Snapper with the probable addition of an infrared homing device for guidance of the terminal phase. Range: 2,500 meters.
Sagger (AT-3)	More compact than Snapper and Swatter, Sagger carries an equally powerful warhead and can be fired from various carriers as well as from the ground. Its firing operation is the same as Snapper's and Swatter's. Reportedly it was quite effective in use against Israeli tanks in the 1973 war. Range: 2,500 meters.
LAW (M-72)	An American light anti-tank weapon, the LAW is a 2.5kg rocket launcher which is replacing the bazooka in many roles. All a soldier has to do is open the launch tube, withdraw the inner tube, raise the sights, aim, and fire. LAW is in Israeli hands. Effective range: 250 meters.
Milan	A French-German advanced portable wire-guided, spin-stabilized missile that may be launched from the ground or from a vehicle. Milan uses a semi-automatic guidance system that only requires the gunner to maintain the crosshairs of his guidance unit on target during the missile's flight. Milan will be made available in large numbers, particularly to Egyptian forces. Range: 25 to 2,000 meters.
TOW	TOW stands for tube-launched, optically-guided, wire-guided. This U.S. missile is a heavy assault weapon system that may be launched from most wheeled or tracked vehicles and from the air at high speeds. Highly maneuverable, TOW missiles can successfully hit moving targets at very short ranges, leaving no visible smoke or heat trails. Its high explosive warhead can pierce the armor of all known types of tanks and can be used against concrete bunkers and other fortifications as well. Like the Milan, the TOW only requires the gunner to maintain the crosshairs on the target during the missile's flight. The Israelis have successfully used both versions. Range: 65 to 3,000 meters.

Table A-14

NAVAL VESSELS IN THE MIDDLE EAST, 1967/68–1975/76

Country/ Vessel Type	1967-68	1973	1974-75	1975-76
EGYPT				
Submarines	8	12	12	12
Submarine chasers	—	12	12	12
Destroyers	8	5	5	5
Escorts	12	4	3	3
Minesweepers and mine countermeasure craft	10	12	12	12
Landing craft	6	14	14	14
Osa-class missile patrol boats	7	12	8	8
Komar-class missile patrol boats	5	7	6	5
Motor torpedo boats	40	36	29	30
Total	96	114	101	101
SYRIA				
Minesweepers	2	3	3	1
Submarine chasers	—	2	—	—
Coastal patrol boats	3	2	2	1
Fast patrol boats	15	—	—	—
Komar- and Osa-class missile patrol boats	—	6	6	6
Motor torpedo boats	—	12	12	11
Total	20	25	23	19
ISRAEL				
Submarines	4	3	2	5
Destroyers	2	1	—	—
Anti-aircraft frigate	1	—	—	—
Coastal escort	1	—	—	—
Landing craft	2	9	9	10
Fast missile patrol boats	—	12+	16	16
Motor torpedo boats	11	9	9	6
Small patrol boats	4	23	30	30
Seaward defense boats	5	—	—	—
Total	30	57+	66	67

Table A-15

MAJOR NEW WEAPONS LIKELY IN THE MIDDLE EAST

Type	Ordnance Cabability (tons)	Approximate Combat Radius (miles)	Avionics, Weaponry and Operational Role
AIRCRAFT			
F-15	7.5	3,450 (ferry range)	A highly maneuverable all-weather air-superiority fighter with a combat tracking ability far superior to that of any other aircraft. The F-15 can be used equally well for air-to-ground missions without sacrificing the success of its primary air superiority role. It carries long-range detection and tracking radar capable of following high-speed targets at all altitudes and adaptable for dog-fights. Loaded with other sophisticated electronic equipment, the F-15 also carries a variety of weapons including four AIM-9L Sidewinder missiles, four AIM-75 Sparrow missiles, and a 20mm gun. Israel is scheduled to receive its first twenty-five F-15s in 1976.
F-16	5.5	575+	An air superiority-ground attack fighter whose chief features are its low cost and a degree of maneuverability that could revolutionize air combat tactics. Its avionics are among the world's best. When it is not armed with nuclear weapons, a typical F-16 ordnance load consists of two-to-six AIM-9 Sidewinder missiles, freefall and electro-optically guided bombs and dispensers, ECM pods, and a 20mm internal gun. Israel expects to receive over 200 as part of the Sinai aid package.
Mirage F-1	4.4	1,000-1,400	A multi-mission fighter and attack aircraft designed as a replacement for the Mirage III. With an internal fuel capacity 45 percent greater than that of the Mirage III, the F-1 has twice the combat radius and thrice the endurance capability for patrol or high-altitude supersonic interception missions. With greatly increased maneuverability compared with its predecessor, the F-1's primary role is that of all-weather interception, although it is equally suitable for attack purposes. Its armament includes two 30mm cannons, two Matra R.530, Super 530, or 550 Magic or two Sidewinder air-to-air missiles, or one AS.37 Martel anti-radar missile, or one AS.30 ASM, and assorted bombs and rockets. Egypt will be receiving at least forty-four F-1s in the near future.

Table A-15 (continued)

Type	Ordnance Capability (tons)	Approximate Combat Radius (miles)	Avionics, Weaponry and Operational Role
AIRCRAFT			
Hawk (Hawker-Siddeley)	Less than 1	1,725 (ferry range)	A two-seat multi-purpose strike/trainer aircraft which is scheduled to be in service by 1976. Egypt will probably procure a large number.
Jaguar	5	500	A tactical support fighter-bomber which can carry Matra Magic or similar air-to-air missiles, Martel AS.37 air-to-surface missiles, and a variety of other ordnance. Egypt is likely to purchase some 200 Jaguars as soon as licensing arrangements are completed.
SURFACE-TO-SURFACE MISSILES			
Pershing 1A[a]			A land-mobile, air-transportable surface-to-surface tactical ballistic U.S. missile. Carried by an erector launcher capable of paved road and cross-country travel, the Pershing has an inertial guidance system and carries a nuclear warhead of 400 kilotons, but is capable of carrying a smaller warhead. Range: 100-500 miles.
Scaleboard[a]			A land-mobile surface-to-surface Soviet missile similar to the Pershing 1A. Scaleboard is believed to have an inertial guidance system and like its roughly equivalent American counterpart was produced for use as a nuclear delivery system. Its warhead has been reported to have a greater destructive capacity than Pershing's. Range: 430-500 miles.

[a] Strong opposition in the fall of 1975 did much to prevent the providing of Pershings to Israel. However, the issue is only shelved, and if the arms race continues, it is possible that new requests may be forthcoming. If the Pershing is given to Israel, the Soviet Union would probably provide Scaleboards to the Arabs.

Table A-16

NUCLEAR-CAPABLE DELIVERY SYSTEMS IN THE MIDDLE EAST

Ground-Launched	Tactical Aircraft[a]
Lance surface-to-surface missiles	F-4 (Phantom)
Scud surface-to-surface missiles	Kfir[b]
Frog-7 surface-to-surface missiles	MiG-21 (Fishbed)
Jericho surface-to-surface missiles	MiG-23 (Flogger)
M-109 155-mm howitzer	Su-7 (Fitter)
M-110 8-inch howitzer	Il-28 (Beagle)
8-inch howitzer	Tu-16 (Badger)

[a] Israel also has so-called "smart bombs," some varieties of which are nuclear-capable and tremendously accurate. They are also highly effective in conventional modes. Yet it is most unlikely that the United States has provided any nuclear arming devices to Israel. This last consideration applies to all weapons itemized in this table: the United States and the U.S.S.R. have no doubt refrained from intentionally supplying the devices necessary to arm these systems for the actual delivery of nuclear weapons. This technology, however, is not beyond the reach of Israel and, perhaps, of some of the Arab states; in case of an emergency, the nuclear devices could be readily supplied and installed by the superpowers.

[b] This new Israeli aircraft probably can carry nuclear weapons.

Source: Robert J. Pranger and Dale R. Tahtinen, *Nuclear Threat In The Middle East* (Washington, D. C.: American Enterprise Institute, 1975); based on data from Table 1, p. 29.

Cover and book design: Pat Taylor